# Units
### and
# Measurement
# Systems

(Second edition)

by
Bruce R. Gilson

2014

# Table of Contents

## Preface.

This book was created because I have seen a lot of material written about its subject matter which just didn't make sense to me. I expect that there will be a lot of people who will not agree with the positions expressed in this book, but they seem to be correct, and it was my feeling that the material in this book needed to be expressed.

I suspect that one reason that I am so conscious of the things which are discussed in this book is my education as a theoretical chemist. For some reason, chemists seem to be the people most interested in "getting the units right." You would expect that *physicists* would be, but while you see many early chapters of chemistry textbooks explaining the rules of unit algebra and the like, it is rare to find such a chapter in a physics textbook. This seems strange to me, because it would seem to matter as much, or even more, in physics as in chemistry.

It may also be true that chemists, especially theoretical chemists, are more conscious than physicists of the fact that the quantities with which they work can be measured in many different ways, because they tend to use more different units to measure the same things. Just in measuring energy alone, you will find chemists having to convert from liter-atmospheres or electron volts to joules or calories as a regular part of their calculations, and specifically in my graduate calculations in theoretical chemistry the conversions between Hartree units, electron volts, cm$^{-1}$, and calories were so frequent that I had to memorize all the appropriate conversion factors.

But whether in work by chemists or by physicists, I have seen such ridiculous statements that I have felt a need to set the record straight, and this is the reason for this book.

I am certain that there are topics that should have been covered in this book but which were omitted; any reader with suggestions should contact me by e-mail at brg1942@gmail.com and the suggestions will be considered for inclusion in a subsequent revised edition.

In the first edition of this book, I discovered a rather

serious error after publication. Nobody has called it to my attention, but it still needed to be fixed. Otherwise, the changes from the first edition are minimal.

Bruce R. Gilson

July 19, 2014

# Chapter 1:   Types of measurements; types of unit definitions.

The theory of measurement, developed by Stanley Smith Stevens in 1946, identifies four different types of measurements (often designated as *levels of measurement:* this is the name under which they are discussed, for example, in the online encyclopedia known as Wikipedia): *nominal, ordinal, interval,* and *ratio* scales of measurement. While it is only the last of these with which this book will be primarily concerned, it is useful to describe them all so that the differences can be understood, and so that the reason for concentrating on ratio scales is clear.

In a *nominal* scale of measurement, the items being measured may be given a numerical code to categorize them, but these numbers are simply the equivalent of names. If, for example, voters' preferences were coded as 1 for Republican, 2 for Democrat, 3 for Libertarian, and 4 for other, it would make no sense to calculate, say, an average political preference of 1.4; as the numbers are simply *labels,* even the order among them is meaningless, and no more can they be used in statistical calculations than if the party names, themselves were used.

An *ordinal* scale provides rather more information. The actual sequential order of the numbers is meaningful, and can be used. The scale commonly used to represent hardness of minerals (Mohs' scale) is a good example of an ordinal scale. If a mineral is designated as 7 on this scale, it is harder than one designated as 5, and anything designated as 6 is certainly in between, but the difference between 5 and 6 is not the same as the difference between 9 and 10, so again such things as an average would not really be meaningful. It should be clear from these discussions that neither nominal nor ordinal scales can be described as having units of measurement, and so there need be no explanation of the reasons for excluding them from this book.

By contrast, *interval* scales *do,* in fact, involve units. However, they are rarely encountered; the only type of measurement that is commonly encountered that is measured this way is *temperature.* (The story of the weather reporter who said, "Yesterday it was 20°; today it is 40°. So today is twice as hot as yesterday" may be a sign of how little many people understand interval scales. As the name "interval" implies, only the *differences* between two temperature readings is meaningful, not the *ratio*, though the ratio of two differences *does* have meaning, as we *can* say that a rise from 20° to 40° is twice as much as a rise from 20° to 30°.) The rarity of the use of interval scales, combined with the fact that it is often possible to derive ratio scales from them (The *thermodynamic temperature scale* defined in the International System of Units is an example of a ratio scale derived from the Celsius scale, an interval scale.), will justify the omission of much consideration of such scales.

And this leaves the *ratio* scales which, as mentioned earlier, constitute the material with which the rest of this book is exclusively concerned. Both interval and ratio scales require the adoption of a unit of measurement. But while interval scales only permit the calculation of *differences* between measurements, ratio scales allow the calculation of *ratios* (which, obviously, is why they are so named). (Ratio scales are sometimes termed *additive* scales, for reasons that will not be gone into here, but because I want to save the word "additive" for another purpose [see Chapter 10], this word will never be used in this book to characterize them.)

At this point it is appropriate to begin the discussion of units of measurement. There are three ways of defining such units. They can be defined in terms of a specific physical object (as was the yard originally), they can be defined in terms of a physical process that will generate the unit (as is the meter today), or they can be defined in terms of other defined units (as is the yard in the United States today; it is officially 0.9144 meter). In addition, one can speak of a fourth process which could be characterized as "implicit definition" of a unit: this will be discussed later in this book.

It is important to note that when a unit is defined in terms of another unit, or if two definitions of units are both in terms of related processes that lead to a fixed relationship between them, there is no way that the relationship could ever be changed as a result of a *measurement*. Since 1 yd = 0.9144 m by definition, if anyone ever tried to do a measurement process that determined the length of a standard yard in terms of the standard meter and arrived at a measurement other than 0.9144, it would not mean that the equivalence of 1 yd = 0.9144 m was wrong; it would simply mean that the objects being compared in the measurement were either incorrect for the yard, incorrect for the meter, or both. While it is probable that most people would understand this for yards and meters, it might be more difficult to understand in other cases.

There was a time, for example, when a meter was defined in terms of a metallic bar, and later a time when it was defined in terms of the wavelength of a specifically defined color of light. And the second was first defined in terms of astronomical phenomena, and ultimately in terms of an atomic clock which used an entirely different standard than the light used for the meter (a standard that is still current). During all these periods, it made sense to measure the velocity of light. There were independently-defined units of length (the meter) and time (the second), and by some process that determined the number of meters that light traveled in a second, this velocity could be measured. But currently, it would be nonsense to measure the velocity of light. The definitions of the meter and second are interrelated; since 1983 the meter has been defined as "the distance traveled by light in a vacuum during a time interval of 1/299 792 458 of a second." By this definition, therefore, it is impossible for the velocity of light to be anything other than 299 792 458 m/sec. (Clearly, this strange number was chosen for continuity; the velocity, as measured in terms of the immediately-previous definition of the meter, was believed to be equal to 299 792 458 m/sec, and the new definition was made to make the new meter as nearly equal to the old as measurements at the time could determine.) It would be just as impossible for the velocity of light to be anything other than 299 792 458 m/sec. as it would be for a yard to equal anything other than 0.9144 m. So any experiment that purported to be a measurement of the velocity of light could only be, in fact, a determination of whether the experimenter's meter and second were equal to the standard ones; and in fact it would either be a measurement of his meter, or of his second, if the experiment had any meaning at all.

At this point it might be appropriate to bring up the subject of one type of unit that has been conceptually treated in two different ways: what is called *capacity,* measured in the customary systems of the United States and Britain in *gallons* and in the countries using metric measurement in *liters* (whether spelled that way or as *litres*). Originally, capacity units tended to be defined either with reference to a standard container (a physical standard) or as the amount of some substance (usually water at a specified temperature) that had a defined weight. It is now thought better to consider capacity and *volume* as equivalent, and to define the units of capacity in the same way as are defined units of volume. For example, in the metric system, the liter was originally the amount of water (at 4° C) that weighed exactly 1 kilogram. By *measurement,* this was determined to be very slightly larger than 0.001 cubic meter; it had been originally intended to be the same, but because the kilogram and meter had been defined by unrelated standards, the relationship between the liter and cubic meter was something that could only be determined by measuring one in terms of the other. In the latter part of the twentieth century, a change in the definition of the liter was made, which from that point forward actually changed the liter/cubic meter ratio from something *measurable* to something *defined* (in the way the speed of light, in SI units, has now been changed, as described previously). A liter is now, *by definition,* 0.001 cubic meter, and no measurement that could ever be performed would ever be able to change this.

As the definition of the SI capacity unit has changed in *time,* the definition of the capacity unit that was standard (but of two rather different sizes) in the United States and Britain varied in *place.* The United States gallon (used only in measuring liquids) has been defined (at least since the nineteenth century, if not since the beginning of the existence of the United States) in terms of the *cubic inch* using the modern concept of capacity as equivalent to volume; a United States gallon is 231 cubic inches, exactly, and never could be measured to be otherwise (This gallon was first adopted in Britain in 1706 as a wine gallon only, but became the United States standard for all liquids). The Imperial gallon, adopted on June 17, 1824 and standard in Britain and many places such as Canada before they adopted the metric system, and legal there for both liquid and dry substances, was defined as the amount of water weighing exactly 10 pounds (at a specified temperature), like the earlier definition of the liter. This means that if one wanted to know the density of water (at least at the standard temperature) in pounds per Imperial gallon (or in any unit related to the pound per any unit related to the gallon, such as ounces per quart), it was not subject to *measurement,* but could be calculated from the definition. Yet in pounds per United States gallon, the density of water (at *any* temperature) is a measurable quantity. (As a result, the ratio between the two gallons was a measurable quantity.)

It should be noted that in the United States the gallon is a unit for measuring the capacity of liquids only, but it was used (at least prior to metrication) for both liquid and dry materials in Britain. In the United States, dry materials are measured in the *bushel* and its subdivisions, but as both the United States gallon and the United States bushel are defined in terms of cubic inches, the ratio of the two is obtainable by definition and is not measurable. (In the United States, there are some subdivisions of the bushel and of the gallon which have the same name: the *pint* and *quart.* They are quite different in size, so when the possibility exists of confusion, one needs to say "liquid quart" or "dry quart.") In Britain, the gallon and bushel were part of the *same system* (a bushel was 8 gallons, as a foot is 12 inches), so their ratio was similarly defined and not measurable. (And in Imperial measurement, "pints" and "quarts" referred to the same size units regardless of whether in liquid or dry measurement.) But just as, in *liquid measure,* the ratio of the United States and Imperial *gallons* could only be determined by measurement, in *dry measure,* the ratio of the United States and Imperial *bushels* could only be determined by measurement. (Actually, prior to the 19th century, the British used a bushel which was defined in terms of the measurements of a container. As a result, its volume could be calculated by *geometric* methods, and this older British bushel was like the United States bushel in being related directly to the cubic inch. But this earlier bushel became obsolete when the Imperial standard was adopted.)

In practice, it makes little difference whether a unit is defined in terms of another unit in the same system (as when a mile is defined as 5280 feet) or in terms of a unit in a different system (as when a yard is defined as 0.9144 meter). It might be that there is a *psychological* difference, but operationally both sorts of definitions are equivalent. On the other hand, there is a big operational distinction between the definition of a unit in terms of another unit (or combination of other units) and its definition in terms of a physical process, because the second type of definition, as mentioned earlier, is subject to measurement. We may not have a precise knowledge of the standard unit of length used by some ancient civilization, because all we have available are the rulers actually used by them, which have been copied from the standard with uncertain accuracy. All we can do is take all these rulers that we have found, measure them in terms of a presently available unit such as the meter, and average them, giving appropriate weighting in the average to each measurement based on our assessment of the precision with which the particular ruler was constructed.

# Chapter 2:   Unit algebra and the nature of conversion factors.

The fundamental fact in measurement is that every measurement is really two things: a number and a unit. The unit is a *standard*, defined in any of the three ways described in the previous chapter. And the measurement refers to how many times the measured object is as large as that standard. So a measurement of "3 feet" means a length 3 times as great as a standard foot, whatever that may be. And since, in algebra, $3x$ means a quantity 3 times as great as whatever $x$ is, we can really treat an expression like "3 feet" by all the same algebraic rules as we use for expressions involving letter quantities. In particular, a quantity in "ft" multiplied by another quantity in "ft" gives a quantity in "ft$^2$," just as $ax$ multiplied by $bx$ gives $abx^2$, and a quantity in "ft" divided by a quantity in "sec" gives a quantity in "ft/sec," just as $ax$ divided by $by$ gives $(a/b)(x/y)$. This leads to a number of secondary units which become defined in terms of the primary units, though many people make statements regarding these secondary unit definitions that fail to realize the complete nature of these definitions. It is, in fact, these incorrect statements that I have seen which prompted the writing of this book.

It will often be convenient to consider only the part of a measurement that consists of the units, multiplied and/or divided as necessary. In particular, two expressions are said to be *dimensionally equivalent* if all the units in the two expressions are the same (including the exponents to which they are raised). Dimensional equivalence does not require equal numeric factors, but one can consider quantities as being of the same kind in a sense, if they are dimensionally equivalent. (However, it will later be seen that in some cases, dimensional equivalence depends on conventions used to define the system of measurement, so that two quantities may be dimensionally equivalent in one system and not in another.)

One can see that if one divides two measurements of the same type, one gets a pure number. For example, 3 ft divided by 2 ft is simply 1.5, with no units attached. This makes sense, since a 3-foot length is 1.5 times a 2-foot length, even if they were measured in different units, so this ratio cannot have any units in it. But if one unit is related to another, being units of the same kind, their quotient also needs to be considered as a pure number. Since 12 in. = 1 ft, for example, the quotient (12 in.)/(1 ft.) has to be considered as equal to the number 1, because any number divided by itself is 1. This property makes the nature of conversion factors clearer. For in arithmetic and algebra, it is a standard fact that the number 1 (one) has two properties that are often used to simplify expressions:

1.  Any number divided by itself is 1.

2.  Any number multiplied by 1 is unchanged.

These are related, of course, because of the fact that multiplication and division are inverse operations, but it is useful to separate them for the purposes of this discussion. Now the important thing about conversion factors is that *they are actually special ways of writing the number 1.* (This is emphasized in some books on chemistry and physics, more often in chemistry books than in physics books, but yet it seems that this fact is not understood by some people.)

It has earlier been mentioned that some units are defined in terms of other units, either within a single system (e. g., a mile is 5280 ft.) or to define a unit in one system in term of another system (e. g., in the United States an inch is legally defined as 2.54 cm.) In either case, we have two things that are in fact equal to each other, so that, by the first of the two properties above, their quotient is 1. So any of the following is equal to 1: (1 mi)/(5280 ft), (5280 ft)/(1 mi), (2.54 cm)/(1 in), and (1 in)/2.54 cm). And by the second of the properties above, it is legitimate to multiply any expression by one of these factors without changing its value. In addition, it is possible to use the rules of algebra to simplify expressions, so that if it is desired to convert, for example, 3 miles to feet, all that is necessary is to set it up is to write:

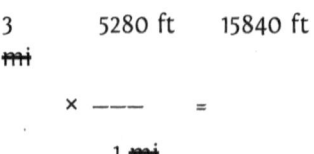

canceling out the miles. Note that if one had tried to use

the factor in the opposite form, as (1 mi)/(5280 ft), the units would fail to cancel, so that using this procedure makes it clear that the 3 has to be *multiplied* by 5280, and not divided. (As noted earlier, this seems to be more carefully pointed out in chemistry books than in physics books, though it would be useful to know in all the quantitative sciences.)

This procedure is applicable, of course, whether the units belong to the same system (as in the above example of feet and miles) or to different systems (as when converting between inches and centimeters using the definition 1 in=2.54 cm); whether the factor comes from a definition or from a measurement (as in a case where one would use a standard bar to define one unit and an atomic standard for the other, or even between two units both of which were defined by standard bars); and even whether the units are primary or secondary (*secondary* units being defined as combinations, such as cm/sec for velocity). Thus if a velocity in ft/sec were to be converted to mi/hr, one would multiply by both the factor (1 mi)/(5280 ft) and the factor (3600 sec)/(1 hr). As all of the conversion factors are expressions equal to 1, any number of them may be successively added to the list of multipliers. For example, if one did not know that 1 mi = 5280 ft, one could use (1 mi)/(1760 yd) and (1 yd)/(3 ft); notice that multiplying them together, canceling units, and collecting numeric factors, the value of (1 mi)/(5280 ft) comes out of that product anyway.

## Chapter 3:   Units in older days.

In the earliest days, there was no coherence to measurement units. Each village might have its own foot, and certainly there was no reason why the same gallon might be used for one liquid as for another. (In fact, in Britain, even into the nineteenth century there were different wine and ale gallons; even now, in the 21$^{st}$ century, petroleum in the U. S. is measured in a 42-gallon barrel, while a barrel of almost any other liquid is 31½ gallons!) Gradually over time, first national governments and later international organizations took over the role of establishing standards for units, though there are still anomalies. (In the U. S., most people are far more familiar with the pound than the kilogram as a measurement unit for weight and mass, even though the pound has been defined in terms of the kilogram since the late 19$^{th}$ century. In Germany, where the pound – Pfund in German – has long been obsolete, McDonald's still sells a *Viertel-Pfunder mit Käse* – a "Quarter-Pounder with Cheese" – though the official weight of the burger in it has been defined in terms of a "Pfund" that is really a half-kilogram, so it is bigger than the same product in the U. S.!)

The history of the standard foot and pound in the United States is interesting. Originally, the United States possessed a "prototype yard" and a "prototype pound," intended to be equal to corresponding British units, and by law all measurements were (at least officially) referred to these prototypes; the standard foot was legally a third of the prototype yard. In 1866, a statute was enacted making the use of metric units legal, and defining the number of inches in a meter as 39.37, with a similarly defined number of pounds in a kilogram. It should be noted that this statute did not eliminate the legality of the prototypes, so that, in effect, what it did in law was to define a United States standard meter and kilogram, which had nothing to do with the international prototypes of those units in Europe; a meter in the United States was 39.37/36 = approximately 1.09361111 times the length of the U. S. prototype yard, and how long it was in terms of the *internationally recognized prototype* was something subject to *measurement*. A similar remark applied to the kilogram. (This should be emphasized: From 1866 on, it became legal to use metric units in the United States, but the metric units were not officially the same units as would be legal in France, because they were based on domestic prototypes. This caused no problems, because the numbers were based on the best measurements anyone had done at that time.)

On April 5, 1893, Thomas C. Mendenhall issued an order that changed this. Mendenhall was the Superintendent of what was then called the Coast Survey, which had control of the Office of Weights and Measures. His order was concurred in by the Secretary of the Treasury, and was conceived of at the time as merely implementing the 1866 statute. (However, although the value of 39.37 inches to a meter conformed to the statute, the Mendenhall Order, by specifying that 1 kilogram = 2.204 622 34 pounds, made a small correction to the statute to implement a better measurement of the ratio of the United States prototype pound to the international prototype kilogram.) Because there was no change in the law, some people have stated that the effect of the Mendenhall Order was simply to redefine the United States standard kilogram. However, in practical terms, the Mendenhall Order is considered to begin the definition of the United States standard units in terms of the international prototypes. (Further changes were made in the exact ratios of foot to meter and pound to kilogram in 1959, in order to harmonize the definitions with other countries. The same new definitions became legal in the United Kingdom and other countries using the customary system of units. The inch is now 2.54 cm exactly, a value that was legal in Canada at the time and slightly smaller than it had been but slightly larger than the previous standard in the United Kingdom; the pound is now 453.592 37 kg, very slightly smaller than the value defined by the Mendenhall Order. Because the topographical maps of the United States Government were all in terms of the old foot, it still has legal standing as the *survey foot*, though the difference between the survey foot and the foot used for all other purposes is so small that most people will never be aware of the difference.)

The importance of international organizations is relatively new, but in almost all scientific work, a system of units known as the International System of Units (customarily abbreviated SI from its French name) is employed (though in some sciences it is supplemented by special units used in each particular field; see Chapters 7 and 8), and many other units customarily employed in the United States and elsewhere are defined in terms of SI units.

# Chapter 4: Convention in measure, part 1: Mathematical formulas.

Most of us are familiar with the formula for the area of a square: $A=s^2$, where $s$ is the side of the square. And we also know that, for a circle, $A=r^2$, where $r$ is the radius of the circle. So it might seem to be a shock when I say that neither is, strictly speaking, correct. The correct statement of the area of a square is $A=ks^2$, where $s$ is the side of the square and $k$ depends on the units employed. The fact is that most of us are so used to the convention that our unit for measuring area is the area of a square whose side is our unit for measuring length that they do not even realize that this *is* a convention. But suppose we were farmers whose common unit of area measurement was an *acre*, while we normally measured distance in *yards* or *furlongs*. Then this would not be so commonplace. The side of a square whose area is one acre cannot be expressed as one square *anything*, whether yard, rod, or furlong. Using any standard distance unit, the side of a square whose area is one acre is some number involving the square root of 10.

Mathematically, one could define an area measurement unit as the area of *any standard shape*, when *some characteristic length* of that shape is equal to our standard length unit. Thus we could actually decide that a circle has the area $A=r^2$, simply by adopting an area unit equal to the area of a *circle whose radius is equal to our length unit*. And this would mean that the area of a square would be given by the formula $A=s^2/$ . In fact, at least at one time, it was customary to give cross-sectional areas of wires in *circular mils*, which were defined in just such a way (though using the *diameter* of the circle, rather than the *radius*). Or we could take as our standard shape as an square, as we usually do, but define our area unit as the area of a square whose *diagonal* was equal to our standard length unit. In that case, we would find that the area of a square would be given by $A=s^2\,2$, and the area of a circle would be given by $A=/_2r^2\,2$.

Yet another possibility would simply be to define our length and area units independently; our length unit could be based on the length of the king's foot, while our area unit could be the area of a field that could be plowed in a farmer's workday (this is not a hypothetical system; it was pretty much the way things *were* measured in the early days!) In that case it would be necessary to resort to *measurement* to determine the value of the constant $k$ in the formula $A=ks^2$ for the area of a square.

The one thing that can be stated with certainty is that, in Euclidean geometry, if we denote by $k_s$ the proportionality factor such that the area of a square is given by $A=k_s s^2$, and we denote by $k_c$ the proportionality factor such that the area of a circle is given by $A=k_c r^2$, then $k_c = k_s$. As we have seen, there is no requirement that $k_s=1$.

One consequence of this approach is that, while it is normally common to read the exponent 2 as "squared" and the exponent 3 as "cubed," this could be confusing. For in the general situation, we are not assuming the convention that (for example) 1 ft$^2$ is the area of a square 1 ft on a side, so calling it a "square foot" or a "foot squared" is misleading. For the purposes of this book, it is recommended that you read these exponents as "to the second power" and "to the third power," instead. As a result, for the purposes of this book, 1 ft$^2$ and 1 square foot are *different* entities: the former is *any unit* obtained by multiplying two foot lengths to produce an area, and thus is undefined until the specific convention (normally the shape such that 1 ft$^2$ is the area when some defined characteristic line in it is 1 ft in length) has been defined, while the latter *specifically* refers to an area unit determined by *assuming* the area of a square is the second power of its side, as a convention. (Recall the earlier reference to "circular mils.")

# Chapter 5: Convention in measure, part 2: Mechanical formulas.

In the previous chapter, the discussion was confined to quantities such as length, area, and volume, where only the laws of mathematics (specifically, Euclidean plane and solid geometry) were concerned. This chapter will go into the types of measurement where the laws of physics enter into the relationships, and it will be clear that *even more* the formulas include special constants that are only determined by convention. In this chapter, the focus will be on units used in one branch of physics, classical mechanics, because the traditional approach has been so much within one convention that nobody ever considers how closely it is tied to that convention. The next chapter will take up the case of units used in electricity and magnetism. This case differs from the case of this chapter, because the conventions used have been recognized explicitly, and as a result, the kind of variation that in this chapter is only offered as hypothetical will be found to have been in actual use.

The usual procedure in mechanics is to set up three primary units, of which two are universally chosen to be units of *length* and *time*. The remaining unit may be a unit of *mass*, in which case the system is what is called an *absolute system*, or a unit of *force*, in which case the system is a *gravitational system*. The three systems most commonly met with are the *meter-kilogram-second* absolute system, the *centimeter-gram-second* absolute system, and the *foot-pound-second* gravitational system; however, all three sets of primary units can be used as a basis for either an absolute or a gravitational system. (See Table 1, below.)

Although the two "metric" systems now considered to be important are the centimeter-gram-second and the meter-kilogram-second systems, historically there have been others. The obvious choice would be the meter-gram-second, using the unprefixed units first defined when the metric system was initially proposed in France in the 18th century. When this system was extended to a fully-blown system, however, it was felt that the meter and gram were really incompatible; the meter was a rather large unit and the gram very small. In the late 19$^{th}$ century, a millimeter-milligram-second system was favored by such people as Gauss, while in 1919, a system based on the meter, tonne (or megagram), and second was adopted in France. The centimeter-gram-second system, however, became the most popular for many years, only gradually yielding to the meter-kilogram-second system. (In 1946, the meter-kilogram-second system was adopted as the primary international standard, with the proviso that it not take effect until 1948. The name "Système Internationale" was formally adopted in 1960, although the SI in fact includes some additional units, so it is not exactly a pure meter-kilogram-second system.)

Before getting into the question of how force is handled in an absolute system, or mass in a gravitational system, a bit of preliminary discussion is needed to characterize two other quantities: velocity and acceleration. Velocity is basically defined as the distance traveled divided by the elapsed time; strictly speaking it is necessary to get into the difference between scalar and vector quantities, but this is irrelevant to the purpose of this discussion and so will be ignored here. In the most usual case, the velocity is changing, and so the processes of the differential calculus are necessary rather than merely dividing the distance by time, but even when the calculus is used the actual process involves dividing a distance by a time and then going to a limiting procedure which, again, is irrelevant to this discussion. So it would appear that velocity is always measured in units that are defined as distance units divided by time units: m/sec, cm/sec, or ft/sec in the three systems earlier mentioned. But this is in fact an oversimplification. One must really appeal to the formula that is generally written $d=rt$, where $d$=distance, $r$=rate (*velocity* in fact being the rate as so symbolized), and $t$=time. And here, in fact, a convention similar to the ones earlier discussed applied. For this equation should be written $d=krt$, where $k$ is a conventional constant almost always taken equal to a pure number 1. In fact, it is always *possible* to choose units of distance, time, and velocity independently, and then an appropriate value of $k$ would have to be determined, probably by measurement, though in some cases by applying mathematical reasoning to known facts. It will turn out that when considering relativity theory, as will be seen in a later chapter, this approach will be followed, but in ordinary "classical" mechanics, nobody would ever consider any convention other than setting $k=1$ in the equation $d=krt$. And so this convention, taking $k$ equal to a pure number 1, *implicitly* defines the unit of velocity; in Chapter 1, this kind of implicit definition was mentioned, and this is one example of its use.

Everything that was just said about velocity being the quotient of distance by time applies, just changing the appropriate words, to *acceleration* as the quotient of velocity by time. The convention is so thoroughly obeyed that it would be inconceivable to most physicists or engineers to measure acceleration in any units but velocity divided by time units, which in turn reduces to distance divided by the second power of time units ($m/sec^2$, $cm/sec^2$, or $ft/sec^2$). Only in the chapter on relativity theory will a different convention be demonstrated, but in theory, distance, velocity, acceleration, and time can all be measured in arbitrarily-chosen units, and the relationships among them worked out by proper choice of the factors in the defining equations.

All the material in the last two paragraphs was necessary before getting into the question of how to define force units in absolute systems and mass units in gravitational systems, because the usual convention is to base the definition of these units on *Newton's second law*, usually written as $F=ma$. When this is done, it is seen that, in an absolute system, a force unit has to be defined as (mass unit)(distance unit)/(time unit)$^2$ (kg $m/sec^2$ or g $cm/sec^2$), and in a gravitational system, a mass unit has to be defined as (force unit)(time unit)$^2$/(distance unit) (lb $sec^2/ft$). Of course, these have been given specific names (respectively *newton, dyne,* and *slug*), but those names are specifically defined as the combinations of units shown here. These are summarized in Table 1.

| System | | Mass unit | Force unit |
|---|---|---|---|
| centimeter-gram-second | absolute | **gram** | dyne |
| | gravitational | (no common name) | **gram** |
| foot-pound-second | absolute | **pound** | poundal |
| | gravitational | slug | **pound** |
| meter-kilogram-second | absolute | **kilogram** | newton |
| | gravitational | metric slug (hyl) | **kilogram** |

*Table 1: Units of force and mass in the most common systems.*

By now, it should be clear that the argument followed in this book is that Newton's second law is inaccurately stated as $F=ma$, and that it ought to be written $F=k_{Nl}\,ma$. (I write $k_{Nl}$, rather than simply $k$, because I want to be able to refer to this specific $k$ at various points in this book.) In fact, although *physicists* and *engineers* regularly use the three systems described here, common usage is to use systems that do not conform to any of these. For the pound, in the U. S., and the gram and kilogram, in the rest of the world, are normally used *both* as units of mass and of weight (which is, in fact, force). And this can only be done consistently if Newton's second law is written with $k_{Nl} \neq 1$. (Most physicists and engineers would say it *cannot* be done consistently; this is nonsense!) If we give $k_{Nl}$ the dimensions of sec²/ft or sec²/m, and the appropriate numeric value, we can treat mass and force as dimensionally equivalent, use the same unit for both, and still have a consistent physics.

How would this be done? The most straightforward way would be, in fact, to use a different equation, rather than Newton's second law, as the defining relationship between mass and weight units. It is well-known that one can use the formula $W=mg$, where $W$=weight, $m$=mass, and $g$ is a local constant (what is meant by local in this context is that it may vary from place to place, but is constant in any specific location). If one simply defines $g=1$ at some standard location, weight and mass are forced to be dimensionally equivalent! (Obviously, the choice of the standard location would make a difference, but for most purposes, anywhere on the earth it would come close to the same relationship. This would really be equivalent to the choice of shape in the previously described length/area formulas, as the dimensionality would not be affected. What it actually measures is the gravitational strength at any location, and this varies little over the surface of the earth.) If $g=1$ by definition, the dimensionality of $k_{Nl}$ has to be sec$^2$/ft or sec$^2$/m, as stated above, to make Newton's second law work. And this, just as Newton's second law, creates an implicit definition of units.

So clearly the usual formulation of Newton's second law as $F=ma$ is not unavoidable, and the ordinary dimensional analysis of force as (mass)(length)/(time)$^2$ is an artifact of the use of that formulation of the law, not a necessity. It can be argued that the second unit system introduced in this chapter, with $g=1$ defined at some standard location, is unsuitable for a general formulation of physics because of its selection of one particular place and giving its gravitational strength the status of a standard. But at this point it will be shown that one can define a system of units that only relies on universal laws and yet does not lead to defining force as (mass)(length)/(time)$^2$.

Consider the law, also presented by Isaac Newton, called the law of universal gravitation: $F=Gm_1m_2/r^2$. Unlike some of the other laws that have been presented so far, this law *does* have a constant appearing in it, $G$. And as this law is usually presented, $G$ is *not* dimensionless, but needs to be given the appropriate units so that it produces a force when multiplied by two masses and divided by the second power of a distance; if we have used the usual convention that force = (mass)(length)/(time)$^2$, $G$ has to have the dimensions of (length)$^3$/(time)$^2$. But the point is that this depends on adopting the usual convention based on Newton's second law having a pure number 1 for $k_{Nl}$. And it is *not* the only way that a system could be established. Suppose that the law of universal gravitation was made a fundamental basis for a system, with $G=1$. This would lead to force having the dimensions of (mass)$^2$/(length)$^2$, and instead of it being necessary to establish the value of $G$ by measurement, it would be necessary to establish the value of $k_{NI}$ in the Newton's second law equation written in the form $F=k_{NI}ma$. Of course, with force being dimensionally equivalent to $(mass)^2/(length)^2$, rather than $(mass)(length)/(time)^2$, many other of the equations of physics would look different from their familiar forms, and the dimensionalities of other quantities would be different from what they are in the usual formulation of mechanics.

It is not the intent of this book here and now to advocate a changeover from the standard formulation of the laws of physics, though it should be clear that this *could* be done. But it will be clear after reading the chapter on electromagnetism that having different formulations which led to different dimensional equivalences of the same quantities is not foreign to physics. It should be clear that what is termed "coherence of unit systems" depends a lot more on convention than is customarily realized. What seems a puzzlement to me is that, given the experience of physics with the electrostatic and electromagnetic units that will be explained in the following chapter, most physicists just cannot conceive that the reliance of the standard formulation on Newton's second law with $k_{NI}$ defined as 1 is purely conventional, and another formulation, such as the ones shown here, could be just as consistent.

One thing that should be noted is that the number of *independently-fixed units* and the number of *convention-setting equations* has a reverse relationship. If one looks at the various quantities that are defined in mechanics based on combinations of length, mass (or force), and time in the conventional formulation that uses Newton's second law with the constant fixed as 1, it was seen that by allowing Newton's second law to contain a constant and not force its use as a convention-setting equation, one could adopt independent units of force and mass. Thus, with one *fewer* convention, one *more* independent unit is possible. (Notice that there are still other conventions, such as those fixing area and volume units as discussed in Chapter 4, or the one fixing a velocity unit in this chapter.) And one could actually use *both* Newton's second law and the law of universal gravitation with the constants fixed as 1, but in that case length, mass, and time units cannot be all chosen independently; choosing a unit for either length or mass will define the other implicitly. (If we require both that $F=ma$ and $F=m_1m_2/r^2$, it can be worked out that an *acceleration,* that is, distance divided by the second power of time, must be dimensionally equivalent to mass divided by the second power of distance, which can be manipulated to make mass dimensionally equivalent to $[\text{distance}]^3/[\text{time}]^2$. This looks strange, but it is a consequence of the adoption of these conventions, since both *distance divided by the second power of time* and *mass divided by the second power of distance,* when multiplied by mass, give a force!) So, with one *more* convention, one *fewer* independent unit is necessary. This is a general rule: each convention set by means of fixing a constant in an equation to 1 (or by fixing the dimensionality of a constant in general) reduces the number of independently-defined units by 1, effectively providing an implicit definition.

It must be noted that setting a constant equal to 1 (as is normally done for $k_{NI}$ in Newton's second law) and fixing the value (*including* units) of some constant (which the current definitions of the SI units effectively do for the velocity of light in a vacuum) are really *equivalent:* they establish a relationship between *units which are no longer independent.* When the second and meter were defined by separate measurements in earlier versions of SI, each definition could be used independently, and the velocity of light could only be determined by measurement. Now, the velocity of light cannot be the object of measurement; it has been *defined.* In exactly the same way, if the value of $k_{NI}$ were to be allowed to be a measurable quantity rather than defined as 1, the units of force and mass could be defined independently. The common popular systems, which use both the pound (or gram) as the unit of *mass* and a similarly-named unit of *force,* would be possible systems to do physics, but $k_{NI}$ might have to be included in a lot of equations where it normally is not today. And one could even decide to measure mass in one arbitrarily-determined unit and force in another (perhaps having a standard piece of matter, such as the prototype kilogram at Paris is now, and a standard force defined with reference to a particular standard spring); with the appropriate value of $k_{NI}$, the laws of physics could still be applied consistently.

# Chapter 6: Convention in measure, part 3: Electrical and magnetic formulas.

While, in the previous chapter, it was shown how quantities of the same type (e. g., force) could be given different dimensional equivalences depending on the conventions used, it was rather hypothetical. Historically, however, in the case of the units discussed in this chapter, they were in actual use in the past, though with the adoption of the International System of Units, they faded from use. There are two different laws of physics that were both adopted as a basis for definitions of electrical and magnetic units: Coulomb's law, which relates the force between two electrostatically charged bodies to the charges on those bodies and the distance between them, and Ampère's law, which relates the force between the magnetic fields produced by electric currents to those currents and the distance between the conductors carrying those currents.

Coulomb's law (in the form in which it will be used in this book) states that if two bodies are charged with electrostatic charges $q_1$ and $q_2$, and the distance between them is $r$, the force between them will be equal to $k_e q_1 q_2 / r^2$, where $k_e$ is a constant depending only on the units employed for measuring force, charge, and distance. In the old *centimeter-gram-second electrostatic system*, the *mechanical* units were defined as previously given for a centimeter-gram-second absolute system, and $k_e$ was set equal to the pure number 1. Given that $r$ was in cm, and force was measured in g cm/sec$^2$, the units of charge (since both charges had to be in the *same* unit!) were cm$^{3/2}$g$^{1/2}$/sec. Because there was a common *practical* unit of charge named the *coulomb* (after the same person whose name was attached to Coulomb's law) the unit of charge in the centimeter-gram-second electrostatic system was commonly referred to as the *statcoulomb*. (In general, names of units in the centimeter-gram-second electrostatic system were given names with the prefix "stat-" attached to the corresponding practical unit. Some texts avoided the "stat-" names and simply referred to the "cgs esu of charge." The name "franklin," in honor of Benjamin Franklin, has also been used for the statcoulomb, but was not common.)

Ampère's law (in the form in which it will be used in this book) states that if two parallel current-carrying conductors of infinite length carry currents $I_1$ and $I_2$ and are separated by a distance $r$, the magnetic force between segments of length $l$ of these conductors will be equal to $k_m I_1 I_2 l/r$, again noting that $k_m$ is a constant depending only on the units employed for measuring force, current, and distance. In the old *centimeter-gram-second electromagnetic system*, the *mechanical* units were defined as previously given for a centimeter-gram-second absolute system, and $k_m$ was set equal to the pure number 2. (There were physical reasons for using 2 rather than 1.) Working out the dimensions in the same way that was done for the electrostatic system, the centimeter-gram-second electromagnetic unit of current, the *abampere*, is determined to be equal to the $cm^{1/2}g^{1/2}/sec$. (In the same way as the prefix "stat-" was attached to the names of practical units to produce the units of the electrostatic system, the prefix "ab-" was used in the electromagnetic system. While it seems obvious that "stat-" comes from the word "electro*stat*ic," the author confesses ignorance as to the origin of the prefix "ab-." One book I have seen says it comes from the word "*ab*solute," but an abampere is certainly not the same as an absolute ampere.) Since current is simply charge per unit of time, the *abcoulomb* would have dimensions of $cm^{1/2}g^{1/2}$. (Again, as mentioned for the electrostatic system, some texts avoided the "ab-" names and simply referred to the "cgs emu of charge." The name "biot" has also been used for the abampere, but was not common.)

The practical units, such as the coulomb, ampere, etc., were in fact fixed so that each was, in theory, a power of 10 multiplied by the corresponding electromagnetic unit. They differed, originally, by small amounts from the exact power-of-ten multiples of the electromagnetic units, because they were defined by methods of measurement more suited to what could typically be done in a 19th century laboratory: the coulomb was the charge whose flow could deposit a defined weight of silver in an electroplating cell, the volt was a particular multiple of the potential developed by a well-defined chemical battery cell, and the ohm was the resistance of a column of mercury of stated dimensions. In time, however, the values were adjusted so that they were the power-of-ten multiples of the electromagnetic units that they had been intended to be.

It can be seen that the two units of charge have *totally different* dimensional equivalences. (In fact, however, the ratio of the two is equal to the velocity of light, which is related to the nature of light as an electromagnetic wave.) And the same is true of *each* electrostatic unit and the corresponding electromagnetic unit. So the hypothetical occurrence which was found of two dimensionally-nonequivalent units for the same quantity, arising simply from the choice of the convention as to which physical law would be used to define a unit, was not just hypothetical but real in electromagnetism. Both sets of units were actually in common usage by physicists, and in fact both would occur in the same physicists' work.

To eliminate this double system of units, two different approaches were followed. In the first, known as the *Gaussian centimeter-gram-second system* (named for the mathematician/physicist Carl F. Gauss), the quantities most closely associated with electrostatics were measured in the same units as in the electrostatic system, while the quantities most closely associated with electromagnetics were measured in the same units as in the electromagnetic system. In the formulation of Ampère's law, the constant $k_m$ was set, not to 2 as in the electromagnetic system, but to $2/c^2$, and a number of other equations used in electrical and magnetic calculations had $c$ or $c^2$ incorporated in them.

All of these three centimeter-gram-second systems are now essentially obsolete. A different approach was taken when the International System of Units was defined. In this system, the mechanical units are those of the meter-kilogram-second absolute system, but neither Coulomb's nor Ampère's law was used to define the units to be employed in electrical or magnetic calculations. Instead, both equations were used in the more complete formulations suggested here, with the constant $k_e$ written in the form $1/4\pi\varepsilon_0$ and the constant $k_m$ written in the form $\mu_0/2\pi$. (The reason for using $2\pi$ rather than the $4\pi$ which would make them symmetric is the same as the reason for using the 2 in other formulations of Ampère's law.) In addition, the practical units, especially the ampere, were incorporated into the International System, with the ampere considered a fundamental unit on an equal basis to the meter, kilogram, and second, and the other practical units defined in terms of the ampere and other meter-kilogram-second units. As a result, the constants $\varepsilon_0$ and $\mu_0$ take on whatever values are necessary to make the ampere have its accepted value. (Because the SI units are simply power-of-10 multiples of the centimeter-gram-second electromagnetic units, $\mu_0$ is numerically equal to $4\pi\times10^{-7}$, but has the units of newton/ampere$^2$.) This is the system currently employed by all physicists and engineers.

## Chapter 7: Nonstandard conventions, part 1: Relativity.

In relativity theory, distance and time are considered to be quantities of the same kind. Therefore, by an argument such as that of Chapter 2, a velocity, which is the quotient of a distance by a time, should be a pure number. Thus it is frequently convenient to adopt a different convention for setting the relationships among units. The most important velocity in relativity theory is $c$, the velocity of light (in a vacuum). So it is made equal to the number 1, like the $k_{NI}$ of Newton's second law in the usual formulation. In consequence, Einstein's famous equation $E=mc^2$, of great importance in relativity theory, becomes simply $E=m$. This requires energy and mass to be expressed in the same units. (This is an example of the phenomenon earlier alluded to, in which quantities that are not dimensionally equivalent in one system might be in another.)

This particular convention, in which energy units can be used to measure mass, is actually often combined, in particle physics, with a particular non-SI energy unit, the electron volt. The electron volt is defined as the energy unit obtained when the charge of an electron is combined with the (standard SI) unit of electrostatic potential, the volt. Since energy and mass are expressed in the same unit under this convention, the electron volt becomes the unit of choice for measuring masses as well. (Actually, its size is such that the masses of particles encountered in physics are expressed in terms of mega-electron volts, symbolized MeV, equal to a million electron volts.)

# Chapter 8: Nonstandard conventions, part 2: Theoretical chemistry/spectroscopy.

Because the author of this book was educated as a theoretical chemist, these particular nonstandard conventions are extremely familiar, so they will be discussed in considerable detail in this chapter. Just as in the previous chapter, an equation (or really two, in this case, but combined into 1) that is important in the discipline in question is used and a constant in that equation is set to 1 rather than using Newton's second law. In this case, the two equations considered as basic are Planck's law, $E=h\nu$, and the general law relating wavelength and frequency, $\lambda=V/\nu$. In the first of these laws, $E$ represents energy, $\nu$ frequency, and $h$ is a constant, named for Max Planck. The second of these equations applies to any wave, and if one is dealing with light (or any electromagnetic wave such as the ultraviolet or infrared radiation with which spectroscopists are also concerned), for the velocity $V$ we can put in the velocity $c$ of light. The wavelength is denoted by $\lambda$ in this equation. Combining the two equations (since the frequency $\nu$ appears in both) we have $E=hc/\lambda$. As with $c$ in the previous chapter, it is convenient to set $hc=1$ when doing theoretical calculations of spectra, so that energies have the dimensions of reciprocal lengths, another example of system-dependent dimensional equivalence. This system is normally used only in describing spectroscopic energies; these are quite frequently expressed in $cm^{-1}$.

Another system which is also used by theoretical chemists relates to the fact that their calculations are made to describe the motions of electrons in atoms and molecules. For this purpose, an equation developed by the Austrian physicist Erwin Schrödinger is used. In the Schrödinger equation, both the mass and charge of the electron appear, and in the usage of theoretical chemists it is common to use these as the *units* of mass and charge, and to set Planck's constant $h$ equal to 2 (rather than to 1 as in some of the previously mentioned systems). When this is done, the energy comes out in a unit named for the British physicist D. R. Hartree. (Hartree originally proposed a number of units which were originally described as *Hartree units,* including the units of charge and mass based on those of the electron; the term now seems to be restricted to his energy unit.)

Theoretical chemists have to be quite adept in converting between different systems of units, because not only are standard SI units used as well as the systems just described, but many other special units (including the electron volt mentioned in the preceding chapter) are found useful in the performance of different types of computations, and the results of one computation, using one system, often need to be combined with the results of another computation, using a totally different system.

# Chapter 9:    A single exception.

Over the years, the nature of units has been to become standardized; in medieval times one village might have a different standard foot from the next one, and gradually each nation would standardize its own units, culminating in the nineteenth century with the adoption of international standards. But in one area, things have gone the opposite way: *currency.*

Through the nineteenth century and into the early twentieth, it was taken for granted that a currency unit represented a fixed weight of gold (or occasionally silver). If you had a coin from a foreign country, you could weigh it (though usually you would consult charts prepared by people who knew the weights of the various nations' coins) and determine its value that way, so that, in fact, anyone knew how much to value such a coin, and people were perfectly willing to accept foreign coins in purchases. (Paper currency was another story: since a paper dollar, say, represented only a *promise* to pay a dollar in precious metal, its value was only as much as the issuer could be trusted to come up with the metal on demand.) And so exchange rates were fixed: over a period of decades, for example, a French franc was worth 19.2¢ in United States currency, because the dollar and the franc each represented fixed weights of gold. (This author remembers seeing notebooks with some of these exchange rates printed on them even in the 1950s, when they had ceased to be correct!) Under this system, an exchange rate was something like the conversion factor between pounds and kilograms, a well-defined ratio.

At some point in the early twentieth century, however, it became obvious that many nations were *unable* to maintain the stocks of precious metal that were needed to insure that the currency had a fixed value in terms of weight of the metal; others, which might have been *able* to do so, became *unwilling.* And so, a definition of a currency unit in terms of a fixed weight of some precious metal was replaced by a system in which the value of a currency unit was determined simply by what the market would charge for its goods. (In Communist countries, instead of the market, the government fixed the prices of most or all goods. And even in nominally capitalist countries, some prices were often controlled.) And this led to floating exchange rates. (For some time after the Second World War, a system was in place where countries would buy and sell other countries' currencies to attempt to stabilize those exchange rates. Eventually, even these attempts at partial stabilization ceased.) So the value of one currency in terms of another became something subject to market trading, like the prices of stocks in a stock market.

The result of these changes is an almost permanent tendency toward inflation. When a dollar meant a certain weight of gold, sometimes gold became desirable enough that a merchant might be willing to accept less than he had earlier asked for his product. So prices went up *and* down. Now, it seems that prices rarely go in any direction other than upward. (Some items, such as electronic devices, do become cheaper. But the overall level of prices never goes down.) While there are people pushing for the resumption of precious-metal standards for currency, it seems to this author that the requirement that it set that governments actively protect the value of their currency is one that current governments are unwilling to assume. So it will not happen, despite the advantages that might ensue from this resumption.

# Chapter 10:   Intensive and extensive properties.

Suppose you have a sample of iron. We can measure a lot of properties of the sample; two examples are its mass (or weight; for most purposes they are the same, unless you are going to move it to a place where gravity is stronger or weaker, but the fact is that some people weigh things by actually determine their weight with a spring balance while others really determine the mass with a balancing instrument), and its density (mass-to-volume ratio). But there are significant differences between these two properties. If a sample is uniform, we can measure the *density* of different portions of it and get the same number; if the sample is not uniform, the density can vary from place to place in the sample, and it is meaningful to measure this variation. By contrast, it would not make sense to measure the *mass* of a part of the sample; the only meaningful quantity is the mass of the whole. Additionally, if one has two samples of different densities, and combines them to make a single sample, one does not add the densities; the resulting sample has, if anything, a density that is somewhere *between* the densities of the two partial samples. But in the case of masses, one *does* add them. So density is a measurement which has two important characteristics:

1.  It can vary significantly from place to place within a sample, and this variation can be measured, and

2.  It is *not additive;* that is to say, if two samples are combined, the resulting sample has a density that is *not* equal to the sum of the individual densities.

By contrast, mass is a measurement which has two contrasting characteristics:

1.  It makes no sense to measure it in some place within a sample; only the *entire* sample has a numerical value that has any meaning, and

2.  It is completely *additive;* that is to say, if two samples are combined, the resulting sample has a mass that *is* equal to the sum of the individual masses.

Most measurable quantities (not quite *all*, but certainly *most*) are like one of these two, and so it is important to note that these are used as a way of classifying different types of measurements. A measurable quantity is an *intensive property* if it behaves like density in the two ways described above, and an *extensive property* if it behaves like mass in the two ways described. (Note that "combining" means different things in different measurements. Volumes only add if one places them together in a way that does not cause a chemical reaction, while it does not matter with masses. Lengths only add if the objects are placed end to end, and so on. All three are extensive properties under the definition of this chapter, but the second rule requires three different modes of combination.)

It might be noticed that the example given here of an intensive property, density, was defined as the mass-to-volume ratio of some sample. This is a typical situation in that many *intensive* properties are best defined as *ratios* of *extensive* properties. If one investigates the uniformity and additivity rules given above, this turns out to be a natural thing: the ratio of extensive properties will generally be an intensive property, provided that these properties lend themselves to measurement on the same samples.

As stated above, not all measurable quantities can be characterized as intensive or extensive properties, but the vast majority of the quantities arising in physics, chemistry, and engineering are one or the other, and most of the quantities arising in other walks of life fall into one of these two categories as well. So this differentiation is important enough to warrant the discussion here.

At this point, the discussion is complete. I hope that this book has proved enlightening, and that it has helped you understand some of the issues in measurement which have been incorrectly treated elsewhere. If so, it has been a worthwhile experience writing the book.

# Index